2885 Sanford Ave SW #28631
Grandville, MI 49418

ISBN: 978-0-9851431-2-1

A Teacher's Guide with discussion questions related to artificial intelligence is available at this website: http://www.bradflickinger.com/zr281.html

A big thank you to the students and educators at
The American Community School of Abu Dhabi,
people like Karyn, Lilia, Joya, Yvette and many more.
Also a big thanks to our wives, Monique and Barbara for
supporting our crazy ideas.
- Brad and Peter

—

The American Community School of Abu Dhabi
encourages its teachers to be innovative and to share their
work with others so that we can all go forward together.

This is Zr281, and Zr281 is a special piece of artificial intelligence code.

Artificial intelligence, also called AI, is the next step in technology that will make life simpler and better for everyone.

Zr281 can make decisions really fast because he can learn and remember more information, better than any human.

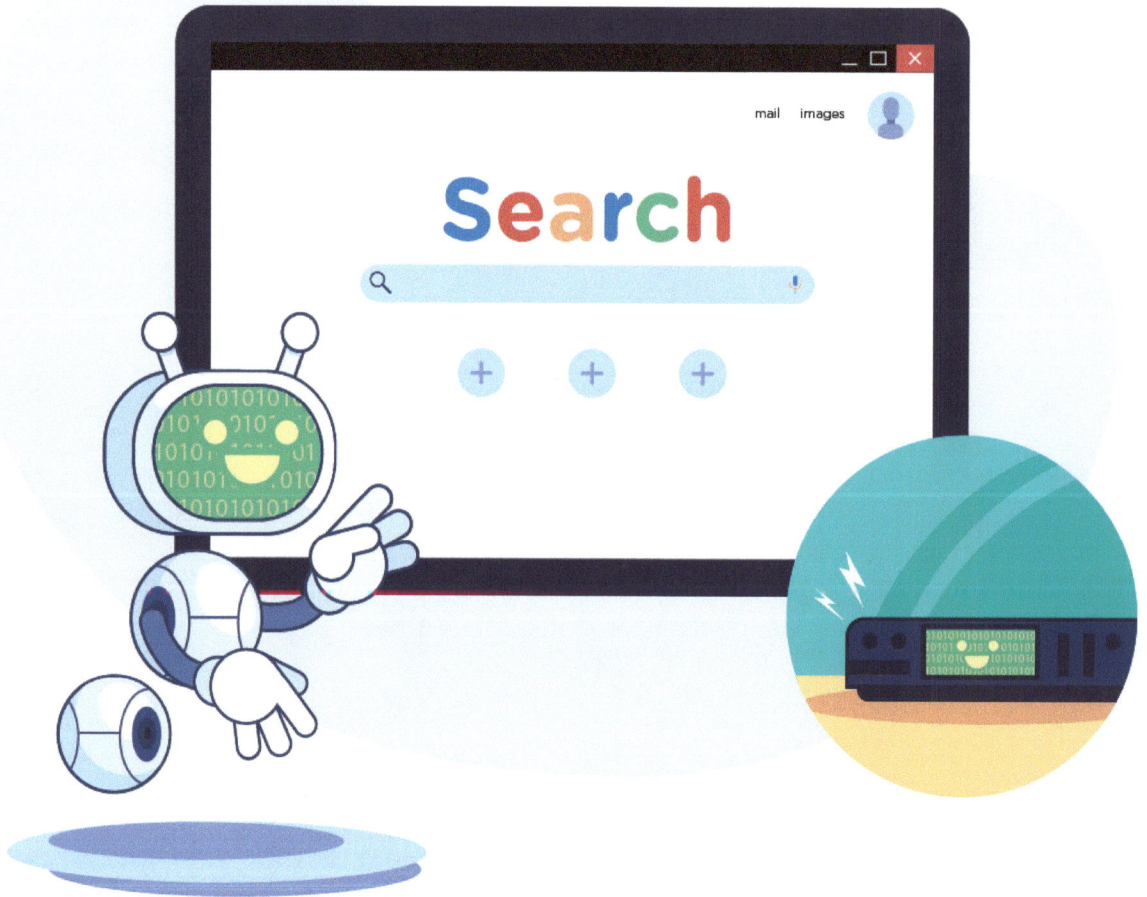

AI is all around us with bots helping us nearly every time we use the internet or a gadget like a robot vacuum at home. AI is designed to make our lives easier and a lot more interesting.

Zr281 was recently created and is looking to find his place in the world. All of his other artificially intelligent friends have already found their jobs for improving human lives.

To try to discover a job for himself, Zr281 decided to spend the day observing his best buddy Gv392, working at her task of operating a self-driving car.

He thought this was such a cool job! She used sensors to operate the car, by paying attention to lines on the road and so much more. After watching how humans drive on the road, she learned not to cross the painted lines and to stop for bicycles.

She was even able to smoothly activate the brakes on a car and avoided hitting a little family of ducks that suddenly waddled across the street. AI's lightning-fast decision-making skills saved the day!

After spending a whole day observing the self-driving car job, Zr281 felt that it was great, but sadly it just didn't feel like his purpose. He decided to continue his search for the job that he was meant to do in this world by talking to his other AI friends.

He went off to visit his other friends working alongside humans in the hospital. There, the AI bots were assisting doctors in diagnosing and treating patients' illnesses. How amazing! Zr281 thought to himself.

His AI friends were able to quickly scan a patient's body using sensors for measuring different levels like temperature, heart rate, and more and in an instant. They helped the doctor give accurate diagnosis of what might be wrong with the patient. This made Zr281 so proud to be an AI bot.

However, Zr281 still had the feeling that this wasn't his purpose in his duty of helping the humans. Over the next few days, he decided to watch a few more friends in their jobs to see if one would be a good fit for him.

He observed an AI bot that was a smart speaker who helped kids finish their homework. These Artificial Intelligence codes used their power of analyzing entire databases, turning voice into words, and then searching in an instant to answer their questions.

Zr281 even visited a great friend at a supermarket. His bot's job was to help people who might forget what they wanted to buy. Unlike other AI bots, humans didn't train him by telling him what the right answer was, instead he observed thousands of shopping carts to figure out what people bought together, like ice cream and chocolate syrup. He uses this to help remind people who bought ice cream that they might want to get chocolate syrup. Or batteries for that new drone!

Zr281's friend shows the strongest shopping combinations to humans who use them, and decides which ones make the most sense. Sometimes it's obvious like ice cream and chocolate syrup while other times it is just coincidence like batteries and baseball bats.

Zr281 went ahead to watch other AI bots working on a farm as drones and weed pullers. These artificial intelligence bots were able to remove weeds that were stealing the nutrients from the delicious crops. Farmers initially trained the AI bots by showing them what weeds looked like compared to the valuable crop. The drone bots would then find the weeds and the weed pulling bots would come and get them.

Zr281's friends explained to him that they were better for the environment and humans. These bots got rid of the need for herbicides to clear the weeds, thereby keeping the fruits and vegetables as fresh as can be. Zr281 cheered on his friend who was the fastest weed pulling bot in all of the field!

But still, he felt that none of these really amazing jobs were what he was meant to do. He started to feel that he just wouldn't be able to fulfill his purpose; to make human life better.

Just then, Zr281 received an assignment request.
His job would be to help blind people function better
in a seeing world.

This sounded so wonderful to him and he thought
maybe, just maybe, he had finally found his purpose.

He first had to go through the data programming process. He would have to learn to be able to see the world just like a human does.

So, he went through supervised training where he had to be shown a picture of an object so many times until he could then recognize that object by himself. At first, he wasn't sure what the difference between an apple and an orange were, but after practicing he slowly began to learn. Not all training processes are the same, but all AI need data training to carry out their duty with accuracy.

After being shown an apple so many times, he learned what an apple was. Then he was shown a picture of a cat so many times, and he learned exactly what a cat was. This process was repeated until Zr281 had learned every object that we know in our world! Just like how the weed bot needed to be shown what a weed was, or the driver bot needed to first see how someone drove, AI's learn from the data humans give them.

Soon after, he was handed his certificate that he had successfully passed his training! He finally felt that he was on his way to fulfill his AI purpose. Zr281 and all of his other bot friends came together to celebrate.

He was placed in a handheld device to make the life of blind people easier. They would simply point this device at objects then Zr281 would identify the object and instantly tell them what it was.

Zr281 was so happy to finally join the other AI bots in fulfilling their purpose of making human life much, much better.

The End

Peter Flickinger started his in eduction helping out at a student help desk. In college he ran a study using A.I. and data from the schools LMS to predict students performance. The simple study proved that teacher have the largest impact on student achievement. and since then Peter has spent time developing programs and trainings to help teachers have streamlined engaging classes.

Brad Flickinger (Peter's dad). has been an educator for over 16 years. For most of that time he has taught elementary and middle school technology classes. Brad designed his tech program so that his students can have a choice to follow their own interests and passions. He believes that the future will be amazing for students who know how to use technology to creatively solve problem.

www.ingramcontent.com/pod-product-compliance
Lightning Source LLC
Chambersburg PA
CBHW041954100426
42812CB00018B/2651